The Power of a Praying Woman

The Power of a Praying Woman

*

The Power of Love, The Power of Hope, The Power of Faith –
That's the Power of a Praying Woman!

Written by: Calvin L. Lewis & Justin O. Cooper

Kingdom Builders
PUBLISHING
CAMDEN, NORTH CAROLINA

The Power of a Praying Woman

The Power of Love, The Power of Hope, The Power of Faith – That's the Power of a Praying Woman!

Kingdom Builders Publishing
an imprint of Kingdom Builders International Ministries

For information address:
Kingdom Builders International
P.O. Box 126
Camden, NC 27921

Unless otherwise indicated, all of the scripture quotations are taken from the *Authorized King James Version* **of the Bible. Scripture quotations marked with NIV are taken from the** *New International Version* **of the Bible. Scripture quotations marked with NASV are taken from the** *New American Standard Version* **of the Bible. Scripture quotations marked with Amplified are taken from the** *Amplified Bible***.**

ISBN 13: 978-1-60141-161-7

Printed in the United States of America

Dedication

Dedicated to & In Memory of:

Betty Jean Shelman, Pearlie Mae Washington, Willie Mae Johnson

Rosetta Chambers, Shirley Ford Sapp, Louisa Thomas, Lillie Mae Robinson, Mother Gladys Hogan, Evelyn Finch, & Vascena Tisdale

You're Legacy Lives On!

Special Dedication

Inspired by & Dedicated to Louisa Shumon, "My Hero"

"Strong Black Man"

By: Calvin Lewis

Who have you been talking to and to whom have you been listening?

Have you forgotten who you are and by whom you came?

You are of a chosen people with strong roots that grow beyond measure Created and shaped in the image of God.

Clothed with Majesty and Crowned with Glory.

You have been given a voice that is able to call those things though they be not, as if they were.

Hands that everything touched by them is Blessed and prosperous.

Eyes that are able to see what is beyond the intellect and comprehension of simple mindedness seeing and discerning the very truth of a thing.

A heart that is sensitive to God and filled with a genuine Love toward your brothers and sisters.

Ears that can separate the very whispers of deceit and untruth and is attentive to the voice of Divine Guidance.
You have a countenance that when you enter into a room you change the very Atmosphere of it and the hearts of all there in and is beyond the intellect of all those that hate on you.

You have a spirit that is beyond what can be discerned in the flesh.

So strong Black Man who have you been talking to and to whom have you been listening?

You may not have the triceps and the bicep's of a body builder or the looks of a super model. But strong

Black Man you have been equipped with something far greater.

You are clothed with a beauty that goes beyond the natural eye and crowned with a Glory that is seen by all that behold you.

So Strong Black Man stop settling for crumbs and buy the whole bakery.

Stop giving up and start looking up.

Remember you are Blessed beyond measure and favored highly.

Walk into your Destiny and claim all that was created for you.

Stop trying to impress and know that accepting you is a benefit on their behalf.

Stop looking for encouragement and encourage yourself.

Stop looking for validation and validate yourself.

Having someone to love you is good but above all Love Your Unique Self.

Yes many try to imitate you that's ok 'cause they sure can't duplicate you.

So get up, shake the dust off and look to what was given to you Strong Black

Man and by that token be the Strong Black Man that you were created to be.

Acknowledgments

Calvin L. Lewis would like to thank:

This book is the result of my heart's desire to salute all the strong women in my life as well as all those who have sacrificed sown tears, torn down barriers and prayed countless prayers so that I am able to reap the benefits from them. I am so fortunate to have been blessed to know strong women and as a result of it I am here today standing as a strong black man.

This book speaks volume because it comes from the heart of my brokenness as well as my victories. I give God all the Glory for the things that he has done in my life. I am thankful for the good as well as the bad knowing that they were needed in order for me to be positioned in the place that I am today. I am thankful to all those who looked at my possibilities and not my inabilities. It has not been an easy journey but thank God I made it.

To my hero, my grandmother, Louisa, who taught me how to love myself and to reach for the stars, I say thank you. Thank you for shaping me into the man that I am today. Despite the many challenges that you faced on a daily basis you unselfishly raised me, not that you had to but because you saw a need to. Thank you for being the Praying Woman that you are. I will always love and appreciate you.

To my mother Berlinda, thank you for making the executive decision to allow me to be nurtured by wisdom and a praying woman. I didn't understand it then, but today I am thankful. I am so proud of you that through all the struggles that you were faced with, you turned out to be a very strong black woman. You are a beautiful and loving person who today I am proud to call mom.

To Justin Cooper, my friend, my publicist, and the co-author of this book, thank you for helping to bring my vision to life. Thank you for pouring countless hours into my dreams to make them a reality. I will always be grateful.

A special thank you to Ellen Newby, my friend, my sister, and prayer partner - thank you for supporting and believing in me. Barbara Hudnall my sister, thank you for being a good role model. My God Mothers, the most powerful praying women I know, Missionary Alver Mae Clary, Missionary Geraldine Peterson, and Mrs. Ruby Douglas Peterson, thank you for praying for me.

Jackie Daniels, thank you for believing in me. Missionary Debra Germany, my Spiritual Mother thank you for shining in my life. Angela Johnson, thank you for pushing me. To all the strong women in my family: Aunt Vera, Aunt Josie, Aunt Jeneva, and Cousin Stella.

A special thank you to DCW staff and all those who prayed with me and for me in the midst of my struggles, I will always love and cherish you.

To my very good friend and brother, Prophet Roderick Evans, thank you for being instrumental in the birthing of my vision. My prayer is that the blessings of God chase you down and consume you. I will forever be grateful to you for what you've done. Thank you!

To Bishop John Gordon and the Covenant Life Deliverance Center (Ft. Lauderdale, FL) thank you for speaking into my life and encouraging me to get to my next level.

And finally, to the man who covers and speaks in and over my life. My Pastor and Spiritual Father, Elder Earl Germany. You are truly my inspiration thank you and I love you!

Justin O. Cooper would like to thank:

In my professional life as a writer penning a book served as the ultimate goal of achievement that I longed to obtain. The Power of a Praying Woman has allowed me in the most fulfilling and most unexpected ways of accomplishing one of my life's goals.

Reflecting on my life and the monumental personal examples of women that have raised me, nurtured me, and inspired me, working on this project was nothing less than a labor of love, gratitude, and therapy for me during a time when my life was in transition.

Father God, the creator of all things, I give thanks to you for depositing the gift of creativity within me, may I multiply the talents you've given me for your glory, may it inspire and encourage others to trust you, love you, and follow you.

Jesus Christ, my personal savior and in whom I can do and accomplish all things, this project yet again highlights that truth that all things are possible through you.

To my comforter and encourager, the Holy Spirit that dwells within me, thank you for pushing me, granting me temperance, self control, and not allowing me to quit.

The depth of my life is measured by the kindness, prayers, love, support, and laughter given to me by so many people that are in the tapestry of my life.

This book is dedicated to the memory of my late grandaunt Evelyn Finch whom I am grateful that our life's path crossed. The foundation of who I am is rooted in your example to always have a good attitude and show others kindness and respect. Your legacy lives on through me. You are truly missed and remembered.

To my mother, Joan D. Hampton, thank you for the many prayers over the years and even now. You are the world's best mother! Your beauty is natural and eternal, internal and external. You are my example of what it means to be used by God. Thank you for your unwavering love, encouragement, respect, and pushing me to aim for the moon. I am thankful for the restoration and the redemption of time that God has allowed us to have. I hope that you are proud of me. I love you!

Ann Nesby – the world knows of you as a singer with an enormous range and emotive intensity, I know you as a friend, you've played a very significant role in my life, you've allowed me to see that dreams are possible and to persevere at all costs. Thank you for all the dinners, exciting road trips touring, and allowing me to use my God given gifts to promote the things of God that you have been called to do. It is an honor having you as a friend and serving you. Thank you!

To Calvin L. Lewis, I thank God for you! Trust is a rare jewel that should be handled with care, thank you for entrusting me with your vision, sharing your hopes and aspirations, and for allowing me to "help you raise what you are giving birth to." You are my inspiration, motivator, and most importantly my friend. Thank you for everything, the best is yet to come and this is just the beginning. Let's Go!

Also Special thanks to: Eric B. A. Jackson, thank you for your prayers, your friendship over the years, seeing me at my lowest, and sharing in my moments of triumph. Ponthetta O. Mitchell, though we don't talk often I am grateful that when we do it's like you never left.

To brothers Trevor A. Cooper and Richard Hampton, Jr, thank you guys for always believing in me no matter how crazy my ideas seemed, I love you both. To my dad Richard Hampton, Sr, thank you for taking care of my mom and being a strong role model not only to me but for so many others.

To my father Thelbert A. Cooper (Barbara) thank you for the lessons of hard work. To my Good Shepherd Outreach Ministry family thank you. Much love to friends and associates professionally and personally, with so many names escaping me know that you are written on my heart. Thank you!

Contents

Introduction

Throughout history women have served as the backbone to the family. The absence of male role models forced many women to take on leadership roles in their homes, churches, and in their communities. The resilience of black women has been paramount. Holding the reigns of single parenthood, these women that have served as the sole breadwinners, the family disciplinarian; it is these women that have cried in the darkness of night finding solace in prayer that "the Lord will make away."

The chronicles of these women written in blood and tears read for us a legacy stretching back throughout the ages and surviving the test of time. The spiritual songs of old that slave mothers would sing as they worked with backs bent towards the earth still resound loudly in the tapestry of our contemporary faith. These women not knowing what tomorrow would hold held fast to the unbreakable faith deposited on the inside of them. It is through their diligent faith and prayer that our heritage has survived.

Contending with the rigid Jim Crow laws of yester year, black women endured the humiliation of having to enter places of business from the rear, the inability to eat at restaurant food counters, to even being told what public water fountain to drink from based on race; it is this group of women that emerged as the grandmothers and great grandmothers of today. It was the grandmother for whatever reason who often took in their grand children when their parents could not or would not provide them with the appropriate care.

Various lessons in manners and respect were taught, "yes sir" and "no ma'am" where the appropriate response when addressing an adult grandma would instruct. The maternal nature of grandma would often perplex the grandfather, but she did whatever was needed going beyond on the call of duty to ensure that her grandchildren were raised up to be respectable, God fearing, and productive members of society.

It is through their tough love, the occasional mint given to you in church on Sunday mornings, by the smell of food cooking on the old cast iron stove, by the many whippings she'd give you to the ones she'd save you from, and most importantly it was through her prayers that we got through.

'The Power of a Praying Woman' is a salute to the legacy of faith filled women that have shaped our lives and taught us lessons love, hope, faith and that prayer changes things.

Chapter One:
The Power of Love

"And now abideth faith, hope, charity, these three; but the greatest of these is love." 1 Corinthians 13:13

*L*ove is that emotion that beyond all things is both complex and simple in its nature. My Grandmother Louisa [Lou –Iza] also affectionately known as my "Nanny" personified for me in demonstration love in action. My mother, who was relatively a child herself when I was born, was not able to provide the care that I needed growing up.

Unlike the many options that mothers have today with adoption to even abortion, growing up in the South and being black post the civil rights era the latter options were labeled taboo. My "Nanny" came out of a generation that believed that family came first, family stayed together no matter what.

My mother thought it best for my siblings and I to live with our grandmother, a choice that I am sure was difficult but necessary for our survival. "Nanny" who was at the time in her forties and remarried, was by all accounts stable and the most logical choice.

The biblical story of Moses displayed an example of a mother's difficult choice to give up her parental rights for the ultimate greater good. The love personified through "Nanny" and her choice to take us in when my mother was unable to care for us set an example of what it means to love.

To abruptly change the course of her life, undoubtedly add unseen stressors to her marriage, and take on the responsibility of motherhood all over again, "Nanny," a woman of faith opened her heart and home to us. Love sacrifices for the greater good, "Nanny" did that for us and she has proven many times over to be my personal hero.

What Love Is:

The scripture so eloquently states what love is, it vividly described as patient, never jealous, boastful, proud, or rude. It is not selfish or quick tempered. It does not keep record of the wrongs that others do. It rejoices in the truth, but not evil. Love is always supportive, loyal, hopeful, and trusting. Simply put love never fails. [1 Corinthians 13: 4-8]

The aforementioned description of what love is should be the prerequisite that all relationships are based upon. Serving as mothers, wives, sisters, and friends, it is in these roles that "praying women" help us to establish our ideologies of what compassion and charity are based upon.

Mary the mother of Jesus stands out as the focal point, epitomizing what it means to be a woman, mother, and friend. Sacrificing a normal life to become the mother of the Son of God, Mary became one of the most celebrated mothers in history.

Being poor, young, and female were all traits that made Mary the most unlikely of people called by God in an age where being a woman, placed you at the bottom of the social, political, and religious class. Despite the societal odds being against her God saw that she would be the ideal vessel in which to birth through the savior of humanity, Jesus Christ.

Being undoubtedly the proudest mother in the course of history, Mary's life was based in the spirit of longsuffering. Unlike the challenges that mothers face today Mary had to deal with idea of her beloved child ultimately growing up to be rejected and publically murdered. Mary simply could have said "NO" to what God wanted to do through her, in essence her life had be one of patience, understanding that her son had to be about His Father's business [Luke 2:49].

Even at the impending and ultimate death of her son, Jesus, Mary's belief that God's plan superseded her desire as a mother demonstrated both trust and hope as she stood at the feet of her crucified son. Her love for her son, our savior, remained until the very end, Mary's love never failed.

The self-esteem that Mary possessed undoubtedly allowed her to understand her value and sense of self worth. Her self-esteem despite her social class coupled with her obedience to God allowed her to take hold and give birth to the vision that God had for the salvation of humanity.

The love that women possess is often rooted in sacrificing their emotion for the greater good. As referenced earlier the mother of Moses had to make the ultimate sacrifice to release her son so that he could ultimately "survive," in essence my mother's choice to allow me to live with my "Nanny" was founded in the same premise.

With uncertainty of what the future would hold for her son, Moses' mother had to release him in love with the faith that everything would work out for the best. There was no guarantee that my life would turn out the way that it has and the same uncertain fate was the path that Moses walked.

When matters of the heart are carried out in love, there is no failure. The faith and belief that by releasing what was most cherished, their children, my mother, Mary and the mother of Moses ultimately had to believe that there was no failure in their sacrificial action of love.

An Enduring Love:

The attitudes women of yesteryear took in their marriages is nearly extinct with the emergences of the feminist movement. The generation of women from my grandmother's era worked to support their husbands, whether it meant serving as the housewife to supporting the vision that their husband had; all based in the spirit of loyalty and an enduring love etched in longsuffering.

Reflecting on some of the not so favorable examples of relationships within my family, women often were the victims of emotional and physical abuse. My "Nanny" by all accounts was a woman in love with her husband and dealt with the unwarranted acts of infidelity on his part. Refusing to step aside as the virtuous woman that God called her to be, " Nanny" personified Proverbs 31:12 . It was through her trust in God and her prayers that she survived, her love endured.

As emotional beings women that are fortified with the power of God and the arsenal of prayer were and are able to overcome the challenges that ill suited relationships bring and unexpected disappointments deliver. When a woman loves she loves with her whole heart often expecting nothing in return. I can recall witnessing moments of pain that "Nanny" endured, she opted to honor the vow she made to her husband, her love endured.

By no means am I condoning being anyone's doormat, because the scripture [2 Corinthians 6:14] advises us not to be unequally yoked in our relationships whether it be marriage, dating, or friendship. Real love is based reciprocity, it demonstrates kindness, respect, honesty, and a commitment to making it work - love endures.

Love Starts from Within:

There is an array of variables which affect the emotional well-being of women intrinsically and by external catalysts. With so many driving forces that dictate to women what the standard of beauty is and what love is from TV ads, movies, novels, to popular music which all supply an endless amount of blatant and subtle messages. These messages often influences a false sense self and create an identity crises of sorts for many young women.

Life holds countless stories of young women that lack self esteem adopting negative and destructive behaviors as a way to compensate for a lack of or longing to receive, give, and feel love. The aforementioned influence of the media only perpetuates these harmful behaviors in many young women. Whether the vice is an eating disorder, drug abuse, objecting themselves through the act of sex in attempt to feel desired. Women that take this path eventually discover that the initial pain is compounded when what they are looking for they cannot find in their self imposed addictions.

For whatever reason - the absence of a father figure in the home, emotional or physical abuse, lack of support or proper validation from family, these women if not met with emotional and spiritual intervention potentially never discover their sense of self or develop a standard of what is acceptable and what they will not tolerate. Moreover, it is these women who fall prey to ill suited relationships with men who prey on their weakness or they simply settle, not knowing that God desires his best in all areas of their lives.

The emotional scars that some women bare makeup cannot cover and the damage runs deep. These women often have been repeatedly misused and abused emotionally and or physically, they begin to develop a level of tolerance below the level that is suitable for any human being. If a man truly loves you then he would do everything within his power to honor you, he would not cheat on you, he would not violently put his hands on you; he would speak with words of love and demonstrate in action his love for you.

Sadly for women who have not developed a sense of self, the love that should be on the inside of them doesn't dictate to them that if the man she is in love with doesn't respect her that it's acceptable. Psalm 139:14 eloquently states that "we are fearfully and wonderfully made."

Singer Kelly Price sung it sweetly in a song entitled 'Strong Man' where she proclaims "That God gave his best when He created me," further expressing the sentiment that love, acceptance, value, and appreciation for self are key to loving others.

If not addressed the vicious cycle continues to spin and these women find themselves eventually in an abyss of broken heartedness, bitterness, and the inability to trust anyone attempting to get close to them emotionally. A pastor friend would often say that before the "real one" appears there sometimes would be a series of counterfeits. These "broken women" who lack the self esteem often repeat the cycle of broken relationships becoming numb to the disrespect that these "ill suited relationships" bring.

As stated earlier it is only through emotional and or spiritual intervention that these women can break free from the bondage that low self esteem brings. Psalm 34:18 lets us know that God is close to the broken hearted, the saints of old would say "Jesus is a heart regulator," we see also that God heals the broken hearted and binds up their wounds [Psalms 147: 3].

We see that in the scripture, that Jesus Christ is the ultimate representation of what love is and what love does. When we accept Jesus into hearts it doesn't matter how messed up we are on the inside because we are made strong in our weakness in Christ [2 Corinthians 12: 9].

The true key to discovering a clear view of self-love is or even for women who have a sense of self-love and esteem is that when Jesus Christ is invited into your heart and mind that you expect the best in your relationships. It's reassuring to know that God calls us by our name and not by our shame, so love yourself and know that you deserve the best when it comes to receiving love.

Love in Demonstration: Loving, Living, & Letting Go:

To paraphrase, the prerogative of God dictates that he uses things and ideas we think in our intellect are crazy or farfetched to prove an eternal truth [1 Corinthians 27]. The truth is that He loves us. In the scriptures, we find many examples of God's love for humanity throughout the ages despite our human propensity for repeat offense and failure. Romans 8:38-39 spells out in detail that nothing in the universe can separate us from the love of God.

A maestro in the art of allowing stories and fate to unfold, God in Hosea 3: 1 instructs his prophet Hosea to marry a prostitute by the name of Gomer, this would serve as a parallel to the relationship that God had with his chosen people, the nation of Israel. The relationship between Hosea and Gomer from the start appeared to be fine, after the birth of their first son we see in the text that she changed drastically. Simply put Gomer began to run the streets and blatantly participated in acts of infidelity and defiance against her marriage and the charge by God that Hosea was called to carry out.

Simultaneously and paralleling Hosea's marriage woes was Israel, God's chosen nation. Hosea served as the prophet warning and compelling Israel to remain faithful and committed to God. Despite the love that Hosea had toward Gomer and the love God had for Israel both Gomer and the nation of Israel opted for a life of rebellion, unfaithfulness, and defiance. Hosea 2 showcases in detail the anguish, disappointment, and frustration that Hosea felt as Gomer willingly opted to cheat on him. It is the same sentiment that God felt towards the nation of Israel as they opted to worship idol gods forsaking the blessings and provisions that He provided for them since their exodus from Egypt.

Later in Hosea 3 we see that despite Gomer's undone condition and repeat offenses Hosea's love forgave the trespasses imposed by his adulterous wife. In the same fashion despite the many transgressions made by Israel, God never gave up on his plan for their lives. Love forgives.

In the Lord's Prayer we are instructed in Matthew 6: 14 -15 that we are to forgive others if we want God to forgive us of our wrong doings. Putting forgiveness into perspective it is evident that women who have been hurt in past relationships find difficulty finding closure, releasing bitterness, and at times seem stuck in suspended animation of what happened to them five minutes, five weeks ago, or five years ago. Sisters you need to let it go!

A friend of mine is adamant with the following points:

In order to receive the new blessings of God, you must be willing to release the old things that no longer serve as an enhancing purpose in your life. Life is a forward motion in which we make use of the wide windshield to see where we desire to go and only looking back at the small rearview mirror of where we've been - we can't drive forward while looking in the rearview mirror.

A common mistake women make is by their unwillingness, to forgive they are unable to receive what they need from God - answered prayer, deliverance, and even forgiveness for the transgressions that they have committed against God. Moreover, the inability to forgive places emotional baggage on the backs of women who refuse to simply "let go."

So you were in a relationship that ended in divorce, your husband cheated on you, your parents were never supportive of you as a child, these are all moments in which you must take emotional control and forgive. Your forgiveness is not a sign of condoning the misfortune you experienced at the hands of others, it is your opportunity to release, know that God will heal your heart, and He will deal with the injustice you faced. Forgiveness allows you to have and keep a heart in which love can be given and received.

Therefore, as you forgive and release the baggage of pain and resentment, you positioned yourself to be blessed by God to move on your behalf and enlarge your territory. So love, live, and let go!

Important Points on Love – 7 Days of Affirmation:

DAY 1: And you shall love the Lord your God with all your [mind and] heart and with your entire being and with all your might. (Deuteronomy 6: 5)

What did you today to honor God through love?

Important Points on Love – 7 Days of Affirmation:

DAY 2: How precious is Your steadfast love, O God! The children of men take refuge and put their trust under the shadow of Your wings. (Psalm 36: 7)

What are trusting God for today?

Important Points on Love – 7 Days of Affirmation:

DAY 3: He who does not love has not become acquainted with God [does not and never did know Him], for God is love. 1 John 4: 8

How have you demonstrated love to your neighbors today?

Important Points on Love – 7 Days of Affirmation:

DAY 4: I love those who love me, and those who seek me early and diligently shall find me. (Proverbs 8: 17)

What have you done to seek the face of God on today?

Important Points on Love – 7 Days of Affirmation:

DAY 5: I give you a new commandment: that you should love one another. Just as I have loved you, so you too should love one another. (John 13: 34)

What examples of love have you demonstrated today?

Important Points on Love – 7 Days of Affirmation:

DAY 6: Above all things have intense and unfailing love for one another, for love covers a multitude of sins [forgives and disregards the offenses of others]. (1 Peter 4: 8)

What have you done through the power of love today to forgive someone?

Important Points on Love – 7 Days of Affirmation:

DAY 7: For I am persuaded beyond doubt (am sure) that neither death nor life, nor angels nor principalities, nor things impending and threatening nor things to come, nor powers, nor height nor depth, nor anything else in all creation will be able to separate us from the love of God which is in Christ Jesus our Lord. (Romans 8: 38-39)

How do you feel about God's love for you?

Conclusion:

We've established what love is based on 1 Corinthians 13 – It is described as patient, never jealous, boastful, proud, or rude. It isn't selfish or quick tempered. It doesn't keep record of the wrongs that others do. It rejoices in the truth, but not evil. Love is always supportive, loyal, hopeful, and trusting. Simply put love never fails.

Our ability to love ourselves outside of our temporal, carnal, and intellectual understanding is based on the universal truth that we all need Jesus as Lord of our lives and hearts. By acknowledging our need for him we will begin to fully comprehend and carryout the commission of 1 Corinthians 13. Our love would be based in and emulate the example that Jesus gave through his life on earth and through his ministry and spiritual example.

When we accept the Lord as our savior we are told in the scripture that God heals the broken hearted and binds up their wounds [Psalms 147: 3]. As you develop your walk and begin to communicate with God in prayer, you can be confident that through Jesus, God will hear and honor your prayer [John 14:14]. If you are honest, transparent, and sincere from your heart God also promises in James 5:16 that prayers of the righteous avail much. God is a healer.

To actively receive from God you must be willing to let go, God instructs us to cast all of cares on

Him [Psalm 55:22], all of your cares includes your past failures, mistakes, anxieties about your current situation, and your future. Tied into having your prayers answered you must remove all resentment and bitterness which sprout from harboring an unforgiving spirit/heart. If God has forgiven you shouldn't you forgive others? In order to move to a better tomorrow you must let go, let God, and move forward.

Prayer of Affirmation:

Now unto Him who is able to keep us from falling and to present us faultless before the presence of His glory with exceeding joy. We thank you for the strong praying woman whose tears have served as the balm to a sick and broken world; whose love has broken the chains and fetters of many who were bound. We thank you for continuing to empower them with the willingness and ability to stand in the gap for us; and to be able to send up prayers of love that actually reach heaven.

In Jesus Christ's name,

AMEN

Chapter Two:
The Power of Hope

"And this is the confidence that we have in him, that, if we ask any thing according to his will, he heareth us." 1 John 5:14

***O**nce you choose hope, anything is possible. – Christopher Reeve (Superman).* "Nanny" emerged out of a generation fortunate enough to establish and build their faith on the cornerstone founded by the black church. Church served many purposes, it was a place of refuge from the injustices imposed on blacks during segregation, it served as the pulse of the community, and it served as a constant reminder to love our neighbors as ourselves [Mark 12:31].

Hope in Harsh Times

The planning, strategizing, and execution of the civil rights movement can be traced to these churches. Legendary women ranging from Coretta Scott-King, Rosa Parks, to Myrlie Evers-Williams served as leaders and visionaries of the hope that was birthed in an era when blacks were considered second class citizens. It was their hope in the vision, the mission, the dream, and the hope that God almighty had a better way that aided in the success of the liberty that blacks and other minorities take for granted today.

Acts of defiance based in the thirst for equality gave courage to Rosa Parks who would not be moved figuratively and literally on that faithful afternoon bus ride; sit-ins throughout the Southeast at food counters, peaceful marches throughout Alabama, to the bloodshed and death that many faced all for one simple cause, hope. That hope has allowed us within this lifetime to witness the election of our first African-American president, Barak Obama, the dream of Martin Luther King, Jr. in part has been realized.

Despite this glimmer of hope that has been manifested the spirit of complacency has gradually crept in to attempt to undermine the promise, the dream, and that hope. It is the little foxes that ruined the whole vine [Solomon 2:15], it is without appreciation and self discipline that liberty becomes the poison that suffocates the progressive hope that prayers once yielded.

Generations, post the civil rights era, have reaped the benefits of their predecessors and in the process have in part disrespected their legacy in their quest for self expression and the mentality of "I can do whatever I want." Complacency has bred selfishness, lawlessness, and disenchantment in many of today's youth. The respect that women had in yesteryear has been replaced by the condoning of derogatory words and negative/stereotypical imagery, which is perpetuated through the media, music, and pop culture. The moral erosion that has taken center stage has opened the door for the enemy to slyly maneuver his way into the lives of our young people.

Once an unconscionable act for a young man to wear his shirt out of his pants this has now been superseded with young men sagging their paints as an expression of urban fashion. To young women who purposely give their temples to men for mere moments of sexual gratification, only leaving them emotionally scarred or worse a single mother, where has the hope gone?

Hope for Better Times

1 Peter 1:13 advises us as believers to be upright and morally alert and set our hope on Jesus Christ. It is by standing on that WORD that we can carry out with great power our command to pray without ceasing [1 Thessalonians 5:17] and stand in faith and expectation that the prayers of the righteous avail much [James 5:16].

When hope is married to confession manifestation is birthed. Hope allows celebration to come by realizing that whatever we were standing in anticipation for is already done in Jesus' name we are simply waiting its materialization. In the contemporary English translation of the Bible, Hebrews 11:1-2 advises us that - Faith makes us sure of what we hope for and gives us proof of what we cannot see.

Our ancestor's faith pleased God. I can recall that same hope and faith exercised by "Nanny," she would tell my siblings and I of her deep desire to see us graduate school and go on to become productive adults. It is her hope, faith, and an undoubted prayer that has allowed me be the man I am today, despite my mistakes over the years her prayers and her instruction have kept me grounded in Jesus Christ. I am sure "Nanny's" prayers were deeply rooted in Proverbs 22:6 – to train up a child in the way that they should go, and when he is old he will not depart from it. The hope that praying woman had has made all the difference in my life.

Important Points on Hope – 7 Days of Affirmation:

DAY 1: You are my hiding place and my shield; I hope in Your word. (Psalm 119: 114)

What is your hope for life?

____

____

____

____

____

____

____

____

____

____

____

____

____

Important Points on Hope – 7 Days of Affirmation:

DAY 2: Be strong and let your heart take courage, all you who wait for and hope for and expect the Lord! (Psalm 31: 24)

What are you waiting on the Lord for?

Important Points on Hope – 7 Days of Affirmation:

DAY 3: And now, Lord, what do I wait for and expect? My hope and expectation are in You. (Psalm 39: 7)

What are expecting today?

Important Points on Hope – 7 Days of Affirmation:

DAY 4: The Lord is my portion or share, says my living being (my inner self); therefore will I hope in Him and wait expectantly for Him. (Lamentations 3: 24)

How has the Lord provided for you today?

Important Points on Hope – 7 Days of Affirmation:

DAY 5: But if we hope for what is still unseen by us, we wait for it with patience and composure. (Romans 8: 25)

What is the most challenging aspect of waiting on the Lord?

Important Points on Hope – 7 Days of Affirmation:

DAY 6: Hope deferred makes the heart sick, but when the desire is fulfilled, it is a tree of life. (Proverbs 13: 12)

What prayer or desire has God fulfilled for you on today?

Important Points on Hope – 7 Days of Affirmation:

DAY 7: But I will hope continually, and will praise You yet more and more. (Psalm 71: 14)

How have you praised God on today?

Conclusion:

Gratitude is the correct attitude when thinking on the power of hope, where it has brought us, how it has kept us, and where it is taking us. Hope is the lantern that we trail after in the darkness of night taking us to our higher calling in Christ Jesus and success in life.

We are that hope, in turn serving as the light of the world through Jesus Christ [Ephesians 5:8] and because He abides in you His word and promises will serve you as a lamp to your feet wherever you go[Psalm 119:105]. So with confidence you can sing "My hope is built on nothing less ... On Christ, the solid rock, I stand; All other ground is sinking sand."

Prayer of Affirmation:

We thank you Father God for the hope that lies within each one of us and the opportunity to reap the benefits of all those that paved the way for us. We thank you for the vision and the provision that you have for them.

This our prayer in Jesus Christ's name,

Amen

Chapter Three:
The Power of Faith

"Now faith is the substance of things hoped for, the evidence of things not seen" –Hebrews 11:1

Praying women all possess one key trait that makes their prayers effective, they have faith. Faith is the fuel that fires the belief that God will turn it around, it is the catalyst in which action is demonstrated toward a desired goal, faith dictates to us our will to go on even when the world around us falling down. Do not ever lose your faith it is your spiritual life force.

There was a woman in my neighborhood like "Nanny" who raised her grandson. This woman was a woman of great faith. As time would go by this woman's grandson began to follow a path contrary to the upbringing that he was taught. This young man took the destructive rebellious path eventually become a statistic (later becoming incarcerated), he became an abuser of the women in his life, to even becoming a denouncer of the existence of God.

Despite the negative reputation that this young man developed, his grandmother always had a positive affirmation when speaking about him. His grandmother in her spiritual wisdom understood that how it looked at the present time wasn't going to be the final masterpiece in which God was orchestrating.

Recently my grandmother told me that she had spoken to this once troubled young man's grandmother and the praise report that she spoke of served as the basis of what it means to "keep the faith." This young man went from being someone that people would purposely avoid, from a life as a prisoner, an abuser, to a societal outcast.

This young man accepted Jesus Christ as his personal savior, joined his local church, and eventually settled down by getting married. This is what I call redemption personified, and it was in part due to the "power of a praying woman," his grandmother, who stood on God's word and the belief that God would hear her midnight cries.

It was her positive affirmation and her confession that began in the supernatural and manifested in reality in God's perfect time resulting in her grandson's deliverance and new found liberty in Jesus Christ. Faith is the key that unlocks the door to God's provision if we believe, this praying woman believed.

The story of the aforementioned young man is among countless examples of how the power faith executed in prayer an action took many troubled youth off of the road to destruction and unto the path of righteousness. Faith for healing, faith for deliverance, faith is what motivates God to move on our behalf, it is faith that ultimately makes you pleasing to God [Hebrews 11:6].

Faith in Action:

Luke 8:43 -47 tells us of the parable of the woman with the issue of blood, this woman for 12 years suffered an embarrassing and debilitating condition that caused her pain, emotional anguish, and shame. Just imagine with your mind's eye at her frustration of going to see countless doctors, specialists, spending hundreds of dollars, and still unable to come to a conclusion as to what her medical condition was.

For her sake, she heard of a man named Jesus who was in the business of healing the broken and the bleeding. It is my belief that when you've tried everything else to no avail the mustard seed of faith within you will press you to go on a little while longer. Since she had tried everything else, I am sure she felt that she had nothing to lose by placing her mustard seed of faith in the man named Jesus that she heard about.

Luke 8:44 states that this woman of faith pressed her way through the mob of gawkers, onlookers, to even those just hanging out to obtain her breakthrough. Sometimes you have to get pass what others may say or think in order to get your blessing, your breakthrough.

In the verses that follow Jesus recognizes that "someone of a greater faith" than just those mobbing him had taken from Him what they needed, this woman of faith received her healing [Luke 8:45-46]. The significance of this biblical parable speaks to us on the importance that faith plays when we put feet to our faith.

It wasn't enough for her merely to believe that her healing was in the man she heard about named Jesus, the corresponding action of getting up from where she was, making her way despite her ailment, and even having to deal with the crowd that swarmed around Jesus, this woman was determined to get what she came for, an answer to her prayer of faith.

James 2:17 advises us that faith without works or corresponding action with our faith is dead. Whatever you stand in belief for from God for you must get up from where you are mentally (renew your mind) and position yourself to receive whatever need you have fulfillment of.

Putting feet to your faith is essential when you can no longer see yourself remaining in the place you find yourself in. It could be moving past the hurts of the past, recovering from mistakes that you thought were insurmountable, to even removing toxic people and relationships that are more of a liability than enriching your life.

In 2 Kings 7: 3 – 20 we see where four men who suffered with leprosy, a disease considered being the worst of the worst in ages past. Outcasts in their city and forced to live outside the city walls of Samaria these four men were hungry, hopeless, and by all accounts at their wits end. Samaria was under enemy attack from Syria and life "as normal" was far from that, as these leprous men contemplated their position they knew that without a corresponding action to counter against their problem that surely they would die where they stood.

Despite their condition of hopelessness these men exercised a faith that prompted them to rise up and walk. Activating their faith God was able to set in motion the windows of heaven being opened up to meet their need and bless countless others in the process. Your obedience, trust, and faith can not only provide your needs but also allow the overflow to reach those around you in need as well.

In the text we see that as they approached the Syrian's army camp that God allowed the mere footsteps of four men to sound like a large armada frightening the Syrian army causing them to flee [2 Kings 7: 5-8]. The Syrian army out of fear fled leaving behind their food, money, and other material possessions that the lepers discovered fulfilling the need that they had and in turn blessing the Samarian people with an abundance; all stemmed from the action ignited by the faith of a few.

The lesson that can be gleaned from this passage is that if these men who were socially, physically, and economically at the base of the social totem pole could rise up despite how they felt, not surrendering to fear, and being proactive to change their situation then the same power is available to us.

Romans 12: 3 clearly explains to us that God the creator of everything has given to everyone a measure of faith. It is imperative that we water our spiritual garden to grow our faith, that we set goals that will allow us to exercise our faith, that we work our faith by putting feet to what we believe God for; faith is the essential attitude to obtaining the promises of God in your life.

An Unwavering Faith:

Examples of faith abound in the Bible, by the documented accounts found therein, where there was faith you often found women. 2 Kings 4: 8 -37 gives an account of the prophet Elisha and his servant Gehazi and their encounter with a wealthy well to do Shunammite woman. This woman would invite Elisha into her home to have dinner with her and her husband whenever he was in the city. As time went by this woman began to feel that Elisha was a true prophet of God, acting on faith she convinced her husband that they build an adjoining room for the prophet so that he would have a place to stay anytime he would come into the city.

Elisha taking the woman and her husband up on their offer would reside in the room that they built for him. Visiting the city with his servant, Elisha asked his servant to get generous Shunammite woman to inquire of her of what she and her husband would need to reciprocate the kindness that she had extended to him. With Gehazi serving as Elisha's translator he asked the woman if Elisha could make a request for her to the king or to the army commander on her behalf, she responded that all of her needs where met and that her family was within close proximity.

Elisha wanting to do something to show his appreciation for her kindness began to wonder what this woman needed that she didn't have. In conversation Elisha asked his servant for his thoughts on the matter, Gehazi told his master that he knew that the woman didn't have any children and that her husband was very old. Elisha then requested his servant to ask the woman to come to the room again and he spoke over her life, prophesying to her that by the same time the next year she would have a son.

Initially the woman asked Elisha not to promise her something that by all natural accounts seemed impossible, removing her doubts of what it looked like or what the circumstances dictated she believed activating her faith, she became the mother that the prophet of God told her she would be.

Like so many of us that in faith have received blessings, promotions, the things we've been praying for, desiring, and standing in need of life still has a way to throw curveballs at you. By all accounts this Shunammite woman's son grew into an active child then out of the blue and unexpectedly the son became ill and died. Obviously, a woman of a certain level of faith the woman kept and held on to her composure. She instructed her husband that she was going to seek out Elisha; she had faith that God operating through this prophet was going to give back to her the very thing that she received from God, her son.

Taking what she needed for her trek to find Elisha she was steadfast in her faith believing that it was not the end, it was not over until God said it would be, she did not waver in her stance, her faith was all she had. Elisha seeing her approach sent out his servant to meet her, determined that she needed to see the man that had spoken over her life she wasn't concerned with dealing with the "middle man," the servant, she needed to see Elisha.

Throwing herself at his feet in distress as any mother would over the death of their child she told the prophet that she didn't even ask for a son and that she swore to God and by Elisha's life that he would come back to her home and her son would live again.

Undoubtedly moved by this woman's persistence and faith Elisha returned with her and through the magnificent and awesome power of God working through Elisha this mother's son was brought back from the dead. If it had not been for this woman's faith in God she could have easily accepted the loss of her son, her faith on fire would not be quenched, she refused to allow her circumstance waver her faith. Her story is one that we need to emulate, no matter what it looks like keep your faith intact, be persistent with what you know God has promised you, and know that God is motivated to move on your behalf based on your faith.

Important Points on Faith – 7 Days of Affirmation:

DAY 1: But without faith it is impossible to please Him, for he who comes to God must believe that He is, and that He is a rewarder of those who diligently seek Him. (Hebrews 11: 6)

How have you sought God on today?

Important Points on Faith – 7 Days of Affirmation:

DAY 2: So Jesus said to them, "Because of your unbelief; for assuredly, I say to you, if you have faith as a mustard seed, you will say to this mountain, 'Move from here to there,' and it will move; and nothing will be impossible for you. (Matthew 17: 20)

What challenges have you faced and overcame on today?

Important Points on Faith – 7 Days of Affirmation:

DAY 3: Jesus turned around and, seeing her, He said, Take courage, daughter! Your faith has made you well. And at once the woman was restored to health. (Matthew 9: 22)

What has your faith yielded you on today?

Important Points on Faith – 7 Days of Affirmation:

DAY 4: Now the just shall live by faith; but if anyone draws back, my soul has no pleasure in him. (Hebrews 10: 38)

How have you demonstrated faith on today?

Important Points on Faith – 7 Days of Affirmation:

DAY 5: That the genuineness of your faith, being much more precious than gold that perishes, though it is tested by fire, may be found to praise, honor, and glory at the revelation of Jesus Christ. (1 Peter 1: 7)

How has your faith been tested today?

Important Points on Faith – 7 Days of Affirmation:

DAY 6: For in it the righteousness of God is revealed from faith to faith; as it is written, "The just shall live by faith." (Romans 1: 17)

How important does faith play in your day to day life experiences?

Important Points on Faith – 7 Days of Affirmation:

DAY 7: Now faith is the substance of things hoped for, the evidence of things not seen. (Hebrews 11: 1)

What are you believing in faith for on today?

Conclusion:

Faith is the essential tool that we need to operate in and utilize on a daily basis to fully receive the bountiful blessings from God. We must get in the habit of rejoicing in the revelation for it expedites the manifestation, we must believe the word, speak the word, and praise God for honoring his word. The biblical examples of faith, the civil right pioneers, to the testimonies of your sisters and brothers in Christ are all examples in which you can enhance and build on your faith.

Prayer of Affirmation:

Father God we thank you for this day and the many blessings you've bestowed upon us. We thank you for grace, mercy, and your unmerited favor. Thank you for the measure of faith that you've given each of us. Thank you for the ability to allow us to put feet to our faith to accomplish the things that you've deposited deep on the inside of us.

We come against the spirits of doubt, anxiety, fear, insecurity, past hurts, and we thank you for healing our hearts from past failed and broken relationships. We thank you now for restoring the years sown in tears.

This our prayer in Jesus Christ's name,

Amen

Chapter Four:
Profiles of Praying Women

Who can find a virtuous woman? For her price is far above rubies. Proverbs 31:10 (KJV)

The purpose of this book is to pay homage to the women who have decided that despite the odds, challenges, or circumstances that they have endured; their love, hope, and faith would be built on nothing less than the rock of salvation that they received and believe upon in Christ Jesus.

Being able to withstand setbacks, disappointments, and mistakes the following women have demonstrated that up against all odds with the help of God that their resilience to overcome superseded the challenges that life presented them. The saying is true for these praying women, "life for me ain't been no crystal stairs."

The following testimonials are from three different women whose life paths differ economically and socially, the common traits they share is that of motherhood, overcoming obstacles, and the evidence demonstrated in their lives that prayer does really change things. To give birth not only to their respective children these women also have had to give birth to the vision, goals, and aspirations planted deep inside of them.

Their birthing experience that these women endured, the labor, the pain, and the occasional scream didn't end in the delivery room of the hospital but extended to their spiritual lives as believers in Christ, productive citizens in a society that challenges their existence, to their everyday survival to maintain mental peace and order.

It is their stories, testimonies, and ultimately their ability to P.U.S.H (Pray Until Something Happened) that they were able to make it over to serve as the spiritual leaders, prayer warriors, and examples of how we should pattern our own spiritual and prayer life to be. 1 Chronicles 16: 8 tells us to Praise the LORD and pray in his name and to tell everyone of what he has done.

This is what it means to allow your light shine before men so that they can see what God is doing and has done in your life, so that they would be compelled to get to know God for themselves through surrendering their problems and most importantly their hearts/lives to Jesus Christ [Matthew 5:16].

It was what the Gospel is all about, sharing the good news, and even in times of distress it's comforting to know that God can take what the enemy and life's circumstance meant for evil and turn it around for your good and ultimately his glory. There is great confidence in knowing that no matter what happens or what may try to form against us that God is the ultimate authority and that nothing catches him off guard or by surprise, He is the author and finisher of our faith [Hebrews 12: 2].

May the stories of challenge and triumph of the following women encourage you in your spiritual walk, show love, believe in hope, and keep the faith.

In Introspect: A Praying Woman - Louisa Shumon aka "Nanny" by Calvin L. Lewis:

People often look at you and think that your life is so easy, it's beautiful, and things just have always went splendidly perfect. That could not be further from the truth in anyone's life, life is hard, it is unfair, and it really comes down to your will to live. That will to live can only be realized through the acceptance of a greater force at work; the power of God made available to us through Jesus Christ.

It was that power at work through my grandmother Louisa that has made a significant difference in my life, her prayers saved me from the path of destruction that I was on. Her example of praying without ceasing was what brought me through the various trials I endured that have led me to this point in time.

I was the third eldest of eight children. Grandma Louisa and her husband raised my older sister, older brother, and I. Our mother was young at the time and made the decision to choose our stepfather over us. I could not understand how or why such a decision of this magnitude came to be, and this caused me to believe in my mind that she didn't love us. This led to confusion and turmoil for me.

The way that I viewed the world and responded to others was altered in such a drastic way that my grandmother couldn't understand my behavior. I was depressed, antisocial, and wouldn't eat. I felt so despondent my soul cried out. There was added stress of not knowing who my biological father was, every time I questioned his existence and where he was in relation to me I was met with unanswered questions. I discovered who my father was on the day that he died.

As a teenager I went through a phase where I would run away from home and battled with suicidal thoughts and feelings, I felt so empty, the pain was immensely unbearable. I recall an instance where I left a suicide note on my grandmother's closet door stating that I didn't want to live anymore and that I was going to jump off a local bridge to the river below.

The internal pain and struggle I felt growing up, I look back now in amazement that I didn't lose my mind, it was my grandmother's prayers that I know undoubtedly kept me.

Life after high school proved to be just as challenging if not more for me than my adolescent years. Relocating to Tallahassee, Florida to attend college I made the futile attempt to join the "in-crowd" and inadvertently got mixed up in the wrong crowd. With my world beginning to spin out of control I began drinking excessively which led to drug abuse, later becoming addicted to both. In an air of confusion and misdirection I felt that I was actually coming into place of "belonging."

My misguided attempts of self love and acceptance were short lived in my affiliation with the "wrong crowd," it all hit a proverbial wall for me when my bout with alcoholism landed me in the front page of my town's local paper, I drove my car into a convenient store. The shame that I brought upon myself led to me being prescribed anti-depressants. My life was in shambles but my grandmother's prayers kept me.

Pulling myself from the mire of shame that I had dug for myself I later married a woman who I thought was the "one of my dreams." The relationship lasted 12 years, it was filled with hardships, joy, sorrow, and the challenges that relationships often bring. At some point along the way we both became strangers in our own house, what I thought love was had come to an end and we opted to divorce.

Looking back on that challenging moment in time I realize that I was married to someone who wasn't created by God just for me; she had not been my wife but a substitute, a drug of choice to enable my codependency problem. Once the divorce was finalized I fell into an abyss of depression which led me to two suicide attempts, each time God blocked what the devil and what my clouded mind was trying to do.

Throughout the many years of adversity I know that it was the Grace of God coupled with the prayers of my Nanny that kept me alive. I am so grateful to have been raised by someone of her magnitude, someone who despite the many challenges she faced in her life, the challenges in her own marriage she still faced each day with a smile, trusting in God to be the way maker that He has promised to be.

There were many times that I wanted and tried to give up, I remember not just what she said but her faith in action. Nanny was perfect in the eyes of a hurting little 12 year old boy who felt alone and misunderstood, she provided comfort to a 17 year old young man who was forced to make decisions that seemed unfair, and to a 22 year old man contemplating taking his life and perplexed at why he couldn't pull the gun's trigger.

It was these times that I reflect on the prayers, gentle reassuring touches of love, and her life living in which she demonstrated in love to her family and the world around her. Despite her own visible and hidden pains she still pressed on through humility always with a spirit to serve.

A woman after God's own heart she opened her mouth she spoke wisdom. When she walked she spoke with compassion, exuded grace, and has for me demonstrated love in action and how the power of prayer could even change a man like me. Thank you "Nanny."

In Introspect: A Praying Woman, Joan D. Hampton by Justin O. Cooper:

My mother Joan gave birth to me a relatively young age, I can only imagine the challenges she faced being young, growing up in a rural area, and uncertain where her life would take her. Much of what I've come to know about her has been through stories she's shared with me and experienced I've witnessed first-hand. Now that I am a young man my understanding of life and the complexities found therein has become illuminated, circumstances and situations that once perplexed me I am now able to resolve.

Thank God for effective communication, a repaired and recovered relationship with my mom, the healing power that God offers, and with time and maturity you just learn to get over what happened that you had no control over.

My mother like most young people trying to find their way spent her twenties juggling motherhood and establishing a career for herself. In the unfairness that life often brought her she had to contend with a tumultuous marriage which later ended in a bitter and messy divorce, to losing a very close sister to domestic violence. She dealt with all these challenges before the age of 26. If had not been for her surrendering her life to Jesus Christ I am sure that the enemy and the stress of life she was presented with she wouldn't be the powerful woman of God that she has grown to be today.

After a very bitter divorce from my biological father in which the court granted custody to my younger brother to my mother and I was in the custody of my father, life as I knew it would change drastically. By all accounts learning that your parents were not going to be together proved to be difficult as a child and the ripple effects are felt to this day nearly 20 plus years later. Relocating away from everyone and everything she knew to revamp her life it was my mother's reliance on God to take care of her that allowed her keep her sanity, dignity, and self respect intact.

In the 20 years since the divorce, my mother has gone from being a victim of circumstance and life to a powerful woman of God that not only I am blessed to call mom, but spiritual advisor, and my biggest supporter. It is because of her prayers in part that I am where I am and who I am today.

After graduating high school and with a brief stint at college I felt the need to expand my horizons by pursuing my dreams, nothing and no one was going to stop me. The aspirations of a 20 year old are at times misguided and like most 20 year olds I thought I had it all figured out, I'd spend the next seven years seeing how wrong I was.

Uprooting the life, I had come to know, I relocated to Atlanta to pursue my career in the music industry, in the process I got involved in an emotionally unhealthy relationship. The great thing about mothers is that they for the most part know their children and they can gauge with a degree of accuracy whether or not the person that their child is in a relationship with is good or not.

In my case my mother was not happy with the choice that I made to be in this relationship, a mother's advice goes a long way in preventing heartache. As aforementioned at the time I thought I knew everything there was to know about life and I was in "love" or so I thought.

After the honeymoon stage of the relationship dissipated, things took a very dark turn, love was replaced with jealousy, disrespect, infidelity, and verbal, emotional, and physical violence. It was the prayers of my mother that I know kept me out jail or even being killed or injured during the little over three year course of that relationship took. Her love coupled with the love of God deposited in her spirit went the distance and protected me.

When the relationship ended and other chapters in my life were closing my mother was there with open arms to take me in, I was a broken man in heart, spirit, and finances. At 24 years-old I felt that my life was over. The mistake of diving head first into a relationship without fully knowing who I was and settling for someone that wasn't meant for me had taken a hefty toll on my spiritual and emotional well being, I became severely depressed, I thought that it was the end.

I had been moderately successful in the pursuit of my career in music entertainment but volatility of the relationship had nearly extinguished my desire to move forward with my life. I thought that I was too damaged to be loved again, I thought that my career ambitions were through, I was going to retreat and live my life in the backburner somewhere on the corner of depression and self loathing.

Thank God for a praying mother! Since the divorce from my biological dad, my mother had moved on becoming upwardly mobile, advancing in law enforcement to a successful career in the insurance sector, to most importantly developing her spiritual walk.

She remarried a wonderful man who to this day treats her like the queen that she is to her and her husband launching Good Shepherd Outreach Ministries in Sumter, SC where I had the opportunity to grow spiritually during my "healing" process. Walking away from that unhealthy relationship, I returned back to SC residing with my mother for a brief stint while I attempted regroup.

My mother continued to pray with me, for me, and over me, she would often encourage me to take what I learned and make it work for me. Her prayers and faith in me pushed me back from the edge of depression and hopelessness.

Through her encouragement, financial support, and prayers I reenrolled in college and graduated with honors in 2007. If it had not been for the intercessory prayer that she sacrificed by waking up walking the floor in the middle of the night I am certain that I could not be where I am today. My mother, Joan D. Hampton personifies and leads as an example of what a praying woman is, and to her I say thank you!

In Introspect -A Praying Woman, Ann Nesby by Justin O. Cooper:

(Photo – courtesy of It's Time Child Records, LLC 2009)

Life has a funny way of giving you just what you ask for, truer words have never been spoken, be careful what you wish for because more than likely you'll get it and a little more in the process. Growing up in rural South Carolina I knew that God had something greater for me beyond the brush of forest and dirt roads.

I knew my life was on a collision course with the things I dreamt about and longed for that the world of entertainment presented me through popular music and TV.

Growing up on the farm with my father, as a teenager, I loathed my life and knew I was going to make my "great getaway" and become someone of importance, I just didn't know how and when it would occur but my intrinsic God deposited faith never allowed me to quench both the desire and fire within.

After graduating high school and attending college, briefly I decided to take a leap of faith and relocate to Atlanta, GA. Leaving the life I knew behind, August 2000 would serve as a life changing moment; it would be the turning point where my love of the arts/entertainment, the hope that I held on to for years, and the faith I had in my heart would meet opportunity. That August I had the privilege to meet one of the greatest voices of the music industry; I was introduced to Ann Nesby.

Ann Nesby was the former lead singer of the Grammy award winning soul group The Sounds of Blackness whose music made up the soundtrack of my life in the 1990s. Because of her I was able to be "Optimistic," "I Believed," I could withstand "The Pressure," and know that with God I could "Go All The Way."

All preconceived notions of what I thought she would be like were quickly dismissed, in my mind at the time I thought anyone with a voice as powerful as hers would exude and personify what is referred to as being a diva. As I walked into her rehearsal for an upcoming concert I noticed her passion and compassion when it came to both her music and how she treated those working with her. I recall her being at the time a robust woman with an equally caring heart.

Not long after our initial meeting, I found myself having the opportunity to be in her presence on regular basis, whether it was for a concert or being invited over to dinner at her home, early on I discovered that even in my in my "faith leaping" the principles I came to learn and grow up under through other praying women in my life, prayer, faith, and hope were all the things both demonstrated in action by Ms. Ann.

Not long after establishing our friendship I was able to be invited into the world that at one time I felt was so elusive, the music industry. I often say that establishing trust is vital to the growth and health to any relationship, my intent and the motive I demonstrated were apparent and Ms. Ann allowed me not only to be her friend but to put to use my God given talents to assist her in the vision that God had given to her through her music.

Despite her previous success her faith ultimately was tried when at the scheduled time of the release of her sophomore album the record company she was a part of at the time decided to release her, in my recollection, Ms. Ann's adult contemporary image was not the direction that the record label felt like they could or wanted to promote since it became more about selling sexual imagery than promoting quality music to the masses.

For a period of time, I remember the phone calls began to slow to a crawl and we were not going out on the road as often as we once were, the measure of faith that Ms. Ann sang about and professed was being tested.

I recall incidents were associates and people wrote her off as "done" and suggested that her and her husband/manager just quit and move on to something else. Despite the crushing blow of realizing the success you've been blessed to acquire is undergoing a testing in which you don't know the outcome, it was her faith, prayer, and praise that got her through the tough times. By 2002 Nesby returned to the top of the charts with the hit R&B inspirational single "Put it On Paper," featuring the Reverend Al Green.

The song became such a hit that Ms. Ann began to tour the country and participated in notable events such as the Essence Music Festival in New Orleans, God had made away and in the process quieted the naysayers. Going above and beyond what she could think or even ask God even blessed the song to be nominated for a Grammy award, one of many nominations that would follow, God proved to be faithful and a God that answered prayer.

Life presenting itself like a railroad track to an unknown destination and as life would have it trials came to again test her faith. Ms. Ann suffered from years of high blood pressure and admitting that her battle with weight was a struggle for many years as an adult and within her professional career as a recording artist. A January 2005 health scare changed her life and once again it was her prayers that got her through.

Ms. Ann was a victim of congestive heart and was also was diagnosed with diabetes, without drastic measures to lose weight her life hung in the balance. After six days in intensive care and praying to God for direction, Nesby decided to undergo bariatric surgery to lose the weight.

Shedding over 120 pounds Ms. Ann admits that she hated that she waited until her health began to be effected to take action and responsibility over her health. In addition to an active prayer life the support of her family and friends have aided her in healing, she notes that she is blessed with a second chance at life.

Since her ordeal with both the music industry and challenges in her health, Ms. Ann has maintained to continue to live a healthier lifestyle through her dietary practices and has garnered additional Grammy nominations with her own music, worked on a major motion picture ('The Fighting Temptations, 2003), to appearing on the highly popular TV show American Idol in support of her granddaughter Paris Bennett (a top 5 contestant from season 5).

Her belief in the power of prayer and praising your way through adversity has made all difference in my world and how I approach the challenges that I face. Ann Nesby personifies what it means to be a "praying woman," her life example of perseverance has encouraged me in the dreams that I pursue to this day.

Chapter 5:
The Power of a Praying Woman
"The Conclusion"

As you read the last few words as this book comes to a close it should serve as your new beginning. If your prayer life was nonexistent or on the fringes of ineffectiveness it's reassuring to know that God is a God of second chances, He is a God of mercy, patience, and compassion. He desires to be in relationship with you.

2 Peter 3:9 encourages us to know that God is a "promise keeper," he is not slack or slow in what he promises us. It is His desire that none of us perish but for us to come into repentance and freely accept his love, his gift of salvation through Jesus Christ.

The biblical examples of women, the heroine acts of women of the civil rights era, to the personal accounts of women that you read about all vary in the scope of time and history. Their socioeconomic status and life experiences differed, however the common denominator to their womanhood was based on their collective reliance on God. Their demonstration of faith materialized Jesus' promise that we can endure, achieve, and serve as effective role models and ambassadors, because in Him we can do all things.

As you embark from this point and moment in time, know and bask in the limitless love of God, embrace your unique self, celebrate your life, and utilize one of the greatest tools in your spiritual arsenal – PRAYER!

Continue to Love…

Continue to Hope…

& Keep the Faith!

Pearls of Wisdom for Daily Living

On Relationships:

- Always know who you are, apart from who you're with.
- You should always be transparent with the one you're with.
- Never take up residency in someone else's world. If the two of you love each other you will create your own world. If you take up residency in someone else's world then you will have to abide by their LAWS.
- Never lower your standards to get to someone else's level
- Always love the one who will always be there for you and that is "you".

- Love never needs to be worked on. If the relationship is stressful then you may need to reexamine things.

- It you can't trust them, then you can't trust them with your heart.

- Never walk with anyone that isn't walking in the same direction as you.

- You should always be able to communicate with each other. If there is no communication then there is no relationship.

- Always know your partner on a spiritual level. It is always healthy to spiritually intertwine.

Pearls of Wisdom for Daily Living

On Loving Yourself:

- You are God's reflection, recognize your beauty and accept yourself and His love for you.

- Don't compromise who you are to receive love from anyone that can't accept and love your unique self.

- Knowing who you are will enable you to be real with others and leaves no room for pretentiousness.

- When you love yourself you respect your self – don't allow anyone in your life that repeatedly disrespects you.

- When in love, love the one your with as you love yourself – remain realistic, honest, and fun-loving.

- When you love yourself you do your best to avoid people, situations, and things that are destructive to your spiritual and emotional health.

- Remind yourself that God gave His best when he created you – your concerns are His concerns and He's only a prayer away.

- Don't feel guilty when on occasion you want to treat yourself – a vacation, an extra piece of cake, or just turning off the cell phone to get in tune with your inner-self are needful to your mental health.

- Show love and compassion to others, it helps you feel better about yourself.

- When in relationships don't be afraid to walk away if the love that you need isn't reciprocated. Love doesn't make you weak or anyone's doormat – love is about mutual respect.

Pearls of Wisdom for Daily Living

On Setting Boundaries:

- If you can't express your feelings don't expect or assume that others can read your mind. Don't internalize a situation if it is bothering you, speak up!
- Silence and avoiding conflict does not make the issue go away, communicate respectfully, empathize, and listen.
- Is your behavior reflective of your Christian core beliefs, are you actions pleasing to God, do they honor your significant other, or would they bring you shame?
- People will only do to you what you allow them to, those that are special to you should be held accountable for their actions and behaviors.
- You cannot control the actions of others – you are only responsible for your reaction and attitude.
- Respect and trust are earned and must be in place for any relationship to survive and thrive.
- You must love God and yourself before you are able to rightfully and respectfully love someone else.
- Treat others they way you'd want to be treated.

- Saying "NO" is okay, Don't over extend yourself, others won't know when enough is enough, no your limits!
- Take one day at a time and don't rush, enjoy the journey. If your significant other loves you they will appreciate you, respect you, and honor you both in word and deed.

Pearls of Wisdom for Daily Living

On Codependency:

- Your codependency should reflect your relationship with God and not another person.
- Never love a creation more than you love the Creator.
- A relationship doesn't validate yourself worth.
- Don't look for a relationship to make you whole, If you only willing to give 50 percent of yourself then don't expect for your relationship to work at 100 percent.
- Be supportive of your significant other but have your own goals and aspirations that they can be supportive of when you need them to be.
- Don't allow yourself to be taken advantage of. Love is a gift you share and not an invitation to be misused or be manipulated.

- Expect and require reciprocity – don't allow your efforts and deposited energy into the relationship to be one-sided, love should be demonstrated to you from the one your with.

- Recognize and respect that your significant other has a life outside of the relationship – be supportive, extend trust, and respect their own individually. Know that if you are important to them you'll almost always get first dibs on their attention so relax!

- Realize that God comes first then your significant other. Thank God for allowing them in your life, appreciate them and the love that they extend to you.

- Never become so full of the person that your with that you become empty in the process.

About the Authors:

Calvin L. Lewis is a self proclaimed "dream chaser" and the visionary founder of Dream Chasers Worldwide, Inc, an organization that promotes and facilitates the emotional and spiritual growth for its members while emphasizing the importance of realizing one's God given dream and manifesting it in the world around them for God's glory. As a motivational speaker, Lewis travels the country on the academic and church circuits inspiring others offering an array of self-esteem building and spiritual empowering conferences and workshops.

He is the author of The Birthing of a Vision and When Sex Interrupts Your Vision. Extending his creative reach to the stage, he is the playwright and director of the inspirational stage play The Gift and The Power of a Praying Woman in which this literary project is inspired by. He is a native of Bainbridge, Georgia.

Justin O. Cooper is a lover of all things creative and has spent nearly a decade pursuing his dream and manifesting his destiny. An honors graduate of Benedict College he has worked with a who's who in the worlds of media and music entertainment. An award winning published songwriter he has worked with Jazz saxophonist Tom Scott and Grammy award winning artist/actress Ann Nesby (former lead singer of the Sounds of Blackness and of The Fighting Temptations fame). A gifted wordsmith, Cooper has worked in an array of capacities as journalist, publicist, and editor. He has served as Nesby's biographer and publicist for nearly a decade, his byline has also been featured in an array of regional and national media publications featuring a laundry list of celebrity interviews in the Gospel and secular arenas. He is a native of Kingstree, SC.

Also, check out the following products by Calvin L. Lewis and Justin O. Cooper:

The Birthing of a Vision – the debut book by Author Calvin L. Lewis available at Amazon.com

Purchase the smooth jazz sounds of Tom Scott's album 'New Found Freedom' where you can enjoy the love inspired jazz infused ballad "You Are My Everything" featuring Grammy award winning artists Tom Scott & Ann Nesby co-written by Justin Cooper – available at Amazon.com.